The
INVITATION

Christianity for Men and Women of Science
A Miracle for Our Time

David Meyer

Linda Morabito Meyer

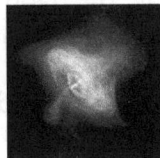

HEAVENS

An Imprint of SciRel Publishing
www.lindamorabito.com

THE INVITATION, CHRISTIANITY FOR MEN AND WOMEN OF SCIENCE, A MIRACLE FOR OUR TIME.

FIRST EDITION

ISBN: 978-0692321102

To Ewald, Cecelia,
Pearl and Howard
who are waiting for us there

~

With our unending thanks

Pastor Brad Viken
Lynda Viken
Pastor Will Glade

"...The advantage of knowledge is that wisdom
preserves the life of him who has it."
Ecclesiastes 7:12

"It was fitting to celebrate and be glad,
for this your brother was dead, and is alive;
he was lost, and is found."
Luke 14:32

Forward

What can this little book accomplish for someone who has rejected Christianity for a plethora of reasons throughout their life?

Consider two scientists, one who has made what is considered the largest discovery of the planetary exploration program for NASA and another who teaches Astronomy to college-aged students, as a retired Air Force pilot... What could they possibly know that can be conveyed in such a short book that could allow a Christian Skeptic with a powerful brain to reconsider Christianity?

Curious? What is behind the content of this book? Is it possible these two scientists don't believe in the science they have contributed to the world and instruct about?

Simply put, what is behind this little book is a breakthrough idea. A very simple premise! Some say Christianity and the Bible can't be real because you can't see science in the Bible. God created the Universe not long ago, and hence the universe is very young. End of story! God created man, and evolution is not part of the Bible. End of story!

But what if God gave us a major clue where the old universe we see today originated, and where and when the process of evolution came into existence? These things are not in the Bible so scientists cannot see the work they do reflected in the Creation Account! Or can they?

Didn't God create the Universe twice? Or to state it another way, didn't God create the Universe once and then change it?

It doesn't say that in the Bible! Or does it?

The Creation Account in the Bible describes a perfect Universe with no death, a Garden of Eden for humanity to flourish in – a Utopian place on Earth to provide us food, beauty, and nourishment for an eternal life on Earth.

Do you live in the Garden of Eden? I wish I did, but I don't think I do! Do you expect to die one day? I expect to, and believe that everyone who is realistic expects to! Maybe the conclusion is that we don't live in the Garden of Eden, as did Adam and Eve! Do you wear clothing to cover your nakedness? Most people do, except in a few places found on Earth, and those places are not the Garden of Eden!

So what happened? God created the Universe exactly as the Bible says, but then He changed it! Is God trying to trick us or lie to us? Absolutely not! God tells us in the Bible that He changed it! The world in which we live, the science that sees an old universe and evolution for all living things, including human beings, is our reality and the universe that exists now, as a result of everything that happened *after* original sin.

So is the earth truly 4.5 billion years old and did cavemen really inhabit the Earth in a complex lineage we call "evolution" several million years ago that has led to our existence? This curious little book contends that **God created the evidence** for both an old universe and an evolved human being, after original sin, after the Universe and everything in it including our Earth became broken by sin.

What else in this little book could open your mind a little about Christianity? Come on! Admit you are curious about what is in this little book! It's short, so that you will read to the end. Could you possibly for the first time in your life find yourself made new, and through sound logic, able to overcome this broken world?

-David Meyer

Contents

Répondez s'il vous plaît

"For many are called, but few are chosen."
Matthew 22:14

Preface

This small book is primarily directed toward scientists, but could serve as a guide to anyone seeking to explore the possibilities of adopting Christianity in their lives.

In the world today there are some renowned scientists who are Christians, and from history as well. A list of their names can be found on some websites on the Internet, including Wikipedia. The reason these scientists are notable is that they are exceptions among the much larger numbers in their profession. There is an ongoing war between science and Christianity, and people have taken sides. The two areas of contention over which scientists and Christians have taken sides to defend their turf are the age of the Universe and how it came into being, and evolution and how life in the Universe began.

This age-old battle between science and religion is not likely to be won on anybody's home court, however. Perhaps the differences between science and Christianity are to be better understood than ever before, rather than overcome. Scientists, who attempt to fit the Bible into science, should not expect to be successful in their approach. The purpose of the Bible is not to verify the advancing scientific theories of any particular time. Christians who try to fit science into the Bible are not likely aware of what science actually is and how it works.

What is profound, however, is that today people who love science seem to be, at a young age, all over this globe, locking themselves into a room and vowing not to come out until they can keep both their love of science and love of their Christian faith in their lives. That means something is happening! Something is happening all over the world!

Let us for one moment conjecture that science is a discipline that when adhered to – its method of investigation and the strides it has brought to the world and the human condition, through discovery and technological advancement – is a truthful and viable way to describe the Universe, to the extent of its comprehension of the Universe so far. Any scientist will indicate that science is a progress report by which theories and models continue to be refined, but overall the laws of nature which have been tested repeatedly have held and are believed to be in effect throughout the Universe.

Now let us propose that Christianity professes the true Word of God, that Jesus Christ is the Son of God, and died for our sins so that we might confess Him, and have eternal life in His paradise.

With the two aforementioned conflicts removed, science and Christianity do not seem to cover the same turf at all or step heavily on one another's feet. When viewed in this perspective, scientists are not markedly different than another professional group we equate with less conflict with the Bible, such as medical doctors. With further consideration, scientists are even more fully appropriately analogous to doctors. It takes a great deal of training and talent to enter either field. Both scientists and doctors improve the human condition and strive to acquire more and more universal insights, so that advancements in their fields can take place.

What then is the result of the conflicts that have caused the war between Christianity and science, if for one moment we step back and take the two preceding paragraphs at face value: the one about how science works and how Christianity's purpose is to reach people with the Good News about Jesus? There is actually only one result. Scientists who are at war with Christianity are rejecting it. There would appear to be few scientists in heaven. (See the aforementioned Wikipedia list for the only ones known to have confessed Christ. It is of course probable that some scientists have not acknowledged their Christianity to other scientists or the public at large.)

Christianity also talks about miracles performed by God and those He sent to spread the Good News about Jesus in the Bible! The authors of this book suggest that perhaps one more miracle has taken place. Perhaps God is missing His scientist children in heaven, and He is asking that they take another look at his Word.

If the source of that war or conflict, the two areas of evolution and the age of the Universe are removed from the picture, how many scientists would be freed from a need to reject the Gospel and the gift of faith from the Holy Spirit? How then might Christianity fit into the worldview of some of the greatest minds on our planet?

10

THE INVITATION

"The Invitation" is written to present a line of reasoning so that scientists do not have to dismiss the God of the Bible, as a starting point. If what contributes to this long-standing battle between Christianity and science is better understood, a great "Reason to Not Believe" might finally stand down. As fellow scientists, we recognize that men and women of science are extremely good at reasoning. Perhaps a very high percentage of those who read this small text completely might think something is drawing them to say, "this makes sense." Perhaps some will say what is presented here is drawing them to believe in Jesus Christ.

WHY THERE IS NO CONFLICT

Christianity and science do not have to nullify each other.
The Bible tells us that all of God's creations were changed
once Adam and Eve defied God.

The Universe we see today is His response to their fall.
Our existence is of course after sin, and science describes
the genetic history Adam possessed,
through evolution, but did not possess originally.

In that same way, the Universe is 6,000 years old,
as in the creation account in Genesis in the Bible.
However, after sin, the Universe, similar to Adam,
possessed a history: **14 billion years of existence**
that God embedded within it.

God then created the light and the history of events
embedded in that light (supernova explosions, etc.) that
left objects too far away from us to reach us within
6000 years (before the universe was created)
as part of how the good and sinless
Earth and everything in and
around it changed.

THE INVITATION

God can do anything!

God created us as curious beings.
We always want to know how things work:
What is over the next hill.
Therefore, God provided us with a Universe
that can make sense to us
as we try to figure out how it works, through science.

What indication of the aforementioned is in the Bible?
When Adam sinned, the Bible tells us that
God changed the very ground of the Earth.
The nature of plants was changed to include
weeds and not just food.
What walks upon the ground, the beasts and human beings
all changed to include the factor of death.
Evolution and an evolving universe,
that had a beginning and will end, are all about death.
These adaptions made by God to His perfect Universe
are what we see today.

Why is this not further called out in the Bible?
It is.
God has left direct indication of this in Genesis.

When the Garden of Eden was planned to be paradise,
God created the birds of the sky before the land animals.
Therefore no indication was given that the birds of the sky
came from the animals of the land,
because they did not,
then.

When Adam and Eve were sent from the Garden,
the changes to the way God had envisioned life for man and
woman on Earth included all living things:
Where birds have the imbedded genetic characteristics
of a land animal called the dinosaur.

The God of the Bible provides one further clue.
Adam was created mature.
His age of maturity had nothing to do with how long it took
God to create Adam.

Here God shares that the Earth, the Universe, and the life
created within it by Him all had histories.
These histories have been uncovered by science.
They are no less important than any of God's creations.
However, they have caused endless turmoil on the Earth.

We are like the blind men all knowing what is there before
us as we try to describe an elephant.
Scientists find the tail.
Christians know that God created the entire elephant,
of course.
But, from where they are standing, there is only
a perfect tusk.

We would deserve the conflict that the Bible only seems to
generate, pitting the creation account in Genesis before sin
against the Universe scientists have uncovered through
science, for not listening to God right from the beginning.

THE INVITATION

It keeps Christians ignorant about what science really is,
and it nearly mandates scientists reject a Universe created
by God, they do not see through investigation:

That Universe and the first man, created by God as
described in Genesis, were given histories exactly as God
has us find today. The History of Adam is found in the
ancient dust of Earth. The history of the Universe is found
through a telescope.

What is science really?

It is God's instrument to protect the Earth from a Near-Earth
Asteroid collision through research that will be used to
deflect the asteroid away from impact with the Earth.
It is the digital imaging technology used in the space
program and by doctors to save the lives of countless
patients around the world everyday.
It is the Nobel Prize in Physics awarded in 2014 for the
development of the LED light source.
Now 1.2 billion of the poorest people on Earth who do not
have electricity, can now own a battery-operated LED
light source for children to do their homework at night.
The battery-operated LED light source costs
less than kerosene in lamps,
that harm children's lungs and cause fires.

The examples of science improving the lives of humanity
are too numerous to chronicle further here.

Science is God's instrument to improve
the lives of His children on Earth.

And perhaps to bring into focus our need for Him.

Why would you want to explore the Bible further?

Read on.

THERE IS NO GOD?

Skeptics who say there is no God often cite the examples from history about the bad things they say have been done in the name of religion.

All the history you may quote to people about what religion has done in the world in your estimation; the things you deem have had the most negative effects on the world have one thing in common.

These things were done by people who may have claimed what they were doing was in the name of God or Christianity.

What part about God giving man free will is incomprehensible to a mind as powerful as yours?

Never confuse what God does and what man does!

It is a trick, a pitfall in your thinking which gives you a reason to not believe.

Are there tragedies in the world such as natural disasters that harm people?

Absolutely!

Do we understand these things, and grasp what God could possibly have in mind when these things happen?

No!

THE INVITATION

Why don't we understand?

Because we are not God!

People who do horrible things in the name of God
are not God.
We also are not God.

Because other people are not God, and because
you are not God,
does that nullify the existence of God?

Of course not!

This thinking would not stand up in any flowchart.
We aren't God, so there is no God?

It separates you from your possible salvation and from
feeling the love that God has for you!

Is it a cop-out to say there are things on Earth we cannot
understand about the Highest Being in the Universe?

Is it a cop-out of any kind to envision a higher order of
intelligence in the Universe that might be so far beyond us,
that there are things about them we would not be able to
fully understand?

Why is this acknowledgement dubbed a cop-out
only when it comes to God?

GOD HAS NOTHING TO DO?

Stephen Hawking has told us that there is no need for a
God, since the Universe did not come with an initial state of
conditions, God would have nothing to do.

Yes, Stephen, the God whose mind you are reading,
as you have reminded us often,
does not exist.

A purely scientific life in current times if lived, places us in an
uncomfortable position that we don't speak about or share.

There are moments in the existence of any scientist
when the very scope of the universe and existence,
in relation to how dwarfed humanity is in
stature and comprehension,
gets through the busy schedule and the research challenges
we embrace,
which rigorously consume our time.

THE INVITATION

I defy any scientist on Earth to deny this happens to them!

What, in addition to the mysteries your science mind is
discerning during the busy hours,
is the Universe telling you during the quiet moments?
When you feel alone, or have trouble,
and feel isolated and so separated from,
unloved, hurt and cheated by the Universe.

This small book is telling you that you now have permission
to know what is missing from your intellectual life
built in science,
and is in no way conflicting with it.

You are invited to not reject Him!
You have always been invited!
God is missing His scientists in heaven.

TEST YOUR INTELLECT

Read the Bible, it's fascinating!

Match your wits with Biblical Scholars
who have been interpreting Scripture for hundreds of years.
(Don't feel bad if the learning curve is steep,
because Biblical Scholars would have trouble
interpreting journal papers in your field!)

They have a head start on you.
Can you keep up in an entirely new field and
hold an intelligent conversation
with your chosen ordained servant of God? (Your Pastor?)

How high is your IQ?
How many times can you read the Bible
and get more out of it each time?

Fill the void.
There is an emptiness that keeps surfacing
in quiet moments.

Can Jesus really fill that void?

Can Jesus really take your problems?

Can you imagine having a personal God
who won't let you down?
Who is always there for you!

THE INVITATION

The Bible is the Word of God.

Jesus is the Word made flesh.

What can you learn and figure out about Him?

You are exploring His Creations.

You are learning about His Universe
with your God-given talent.

Bring forth your prayers to Him.

God hears our prayers!
Enter into a personal relationship with
the Creator of the Universe!
Discuss with the Creator of the Universe your
attempts to bring glory to Him
through understanding the magnificence
and workings of the Universe
He made for us.

THE DISTANCE

There is a great distance between you and your children.

There is a great distance between you and your spouse.

And you live with them in the same house
some if not all of the time.

Does meaningful family time sound good to you?

That aloneness, that separateness is part of life.

Correct?

That is the way I grew up!
There is nothing wrong with my life, you tell yourself.
I am always working sure, but when I do have a break
I need to have my free time the way I want it.

That loneliness or perhaps even boredom, everybody has it.
Everybody needs to deal with it.

Kids find their way in the world by themselves.
They should go through what I went through, you know, all
the successes and failures.
The world is the same as it has always been.

THE INVITATION

Yet I notice my kids are disobedient.
They are materialistic.
They are into what-type-of-activity I don't know
that might actually prevent
them from meeting their dreams and expectations,
as I am living mine?

No way!
Kids always get through.
They won't end up in some kind of trouble, will they?

Hopelessly misguided, or dead by overdose?
No way, I refuse to see it!
I won't look.

Not even into the actual level of happiness or contentedness
of my spouse.

This is a scientist's life and I am living it.
Yet, the world does seem smaller than it ever has.
Information of all kinds reaches our children's ears it seems,
regardless of their ages.
The world seems less sane than it ever used to.

Instant gratification, greed, loss of value of history, integrity,
education, intellectual pursuits, generosity,
lack of caring for one's neighbor
seem to predominate thought.

No, no way, that's not how it is.
It only seems different than at any other time in
civilized history.

What can I do anyway to change this?
It's the same for everybody.

Is there wisdom in the Bible?
Is the wisdom of God available for my perceptions?
Would I make fewer mistakes and guide my family with
strength if I had that wisdom?

Would the activity of us going to church as a family or
praying together
make a difference?

Might this be the best foundation that could be laid for a
family in modern times?
Maybe.

Because I sure would like to lessen that distance
I perceive has developed between my children and me.

I sure would like to lessen the distance
I perceive has developed between my spouse and me.

Where do I start?
How do I start?

WHY DID I DO THAT?

Why did I do that?
Why did I have an affair?

I feel I am entitled to an affair.
Did it hurt my spouse?
Well, probably.

"Why did I do that?" you ask yourself.

Why did I cause so much pain?

Because I am human and a sinner!

You are forgiven.

Your sins are covered in the blood of the Lamb.
Covered in Jesus.

You are forgiven.

Now you too can learn to forgive.

THE COROALLARY TO COSMOLOGY

Cosmology,
the origin and fate of the Universe in modern science
expresses the possible fate of the Universe
in terms of a Big Bang, a Big Crunch, or a Big Rip!

It is an exciting time in science
as scientists seek to uncover the workings of the Universe
at the deepest levels of understanding we can obtain!

What we learn about the Universe, we learn about
ourselves.

This is because *we* are what is out there in space, of course;
the Earth, like any of the objects we study in the Universe,
is a planet of countless trillions in our Universe.

Change in the Universe comes about primarily through
violence.
Observations from the Great Observatories in space have
shown us this.

Consider for the moment, if life survives
any of those processes
such as a Big Bang, a Big Crunch, or a Big Rip?

It does not seem likely.

THE INVITATION

Big Bangs, Big Crunches and Big Rips
are more catastrophic than the way in which
planets are formed, battered, and destroyed
during a proposed period of heavy bombardment
when solar systems form.

Big Bangs, Big Crunches and Big Rips
are more catastrophic than even the way massive stars die
or galaxies collide, cannibalize one another or
gravitationally perturb one another,
as the Universe evolves.

Yet we as human beings seem to care
a great deal about
what will take place long after we are dead.
We take out life insurance policies.
We do research in Cosmology.

What for, though?
What's the point, if the Universe is going to be
destroyed anyway?
Data about the Universe suggests a long-term scenario of
violent death.

What about life?
What was it all for, when any of this is going on?
Life cannot likely survive Big Rips, Big Crunches, Big Bangs!

What was it all for?

I never thought about that before.

The science I love, especially Cosmology,
and love to think about,
suggests everything is destroyed in a cosmic march
toward an ending.

Would I choose to live if I could?
I will die a lot sooner than any of these
end-of-the-Universe scenarios.

What if I could live and no longer be
Earth-bound when I die?
What if I could enter God's perfect paradise when I die?
Would I then be around to see the New Earth
when God creates it?

It is true, embedded in Cosmology,
the ultimate fate of life and the Universe,
is complete destruction.

Therefore death is the scenario of Cosmology.
You know I enjoy reading stories about people
who survive disasters.
If there were an escape vehicle or a
transporter to eternal life
in a sanctuary or perfect paradise of some kind,
would I choose to do that?

THE INVITATION

Why would God create a Universe whose destiny
is to be destroyed?

Is it possible that Christianity which promises eternal life
to those who receive the gift of faith from the Holy Spirit
is not in opposition to science?

Is it possible that Christianity is rather
the logical missing piece of science
and what we observe in the Universe today?

If I could just board a "rescue vehicle"
then I would survive not only my own death,
but survive to see the Big Crunch or even the next Big Bang,

I could see the new Earth, the new Creation, the new Age
the Bible talks about arise!

Could it be that God knew that scientists could present the
evidence for complete destruction of the Universe?

Could it be that God as part of His plan gave the Universe
this history to establish the data for the evidence of this
future, as incentive to seek Him as the only answer?
I certainly can't do myself much good by rejecting Him!

Please, where can I sign up?

THE DEFAULT PARAMETER

Thanks for the information about not rejecting God!
I'll think about it later.

No.
That is not how it works!

The default parameter is outer darkness,
weeping, and gnashing of teeth.
That is Biblical descriptive language for hell
and eternal damnation.

This sounds like superstition or dark fairy tales of old.
It's not.

Right now you are thinking,
you refuse to expose your family to this type of thinking
that got us out of the dark ages in the past!

You will decide how it works, what the truth of the Universe
is with your mind.
That, is the way you have lived up until now,
and the way it always will be!

THE INVITATION

You, in your infinite wisdom, you who sense the distance
between you and your family and the distance between you
and meaningful connection with the Universe, and further
sense, like the expansion of the Universe, that distance is
not only increasing with time, but accelerating.
You know so much, are so smart.
It will be your way.
You know so much that a simple default parameter was all
you had to understand to stop rejecting the Holy Spirit and
to keep you and possibly your family from eternal death.
And you won't do it.
You can't understand the way that it works.
Not understand?
Something this simple?
That could never, never happen to someone
as smart as you!
After all, you know everything.

Don't you?

How difficult would it be to fall on your knees right now
and realize God chooses to bring you to Himself?
God who loves us so much that he gave his only Son to die
for our sins and have eternal life in a perfect paradise.
The Lord embraces us fully within His arms
Our sins are forgiven because He sees us through the
blood of his Son.

You are perfect to Him if you are covered in the blood of
Jesus!

Is it actually okay to humble yourself before
the Creator of the Universe?

Here's how to figure this out.
First, you need to rescale your perspective of who you are,
the pompous vision you have of your importance and your
bombastic view of the extent of your knowledge!
(This applies to all people on Earth, so please don't assume
you are above other people even in condemnation.)

It's okay to renormalize the scale of your importance,
with God,
the Creator of all the Universe,
a little above you.

It's okay to let go of your overall superiority in this world as
you realize you shall one day lie dead in a box or a pile of
ashes in an urn, without your fancy possessions,
words, thoughts, or deeds

Let it leave your hands right now, and lighten your load,
and lift this heavy burden of your life
which at times seem overwhelming.
Let it become weightless.
Release this foolish pride of yours!
Let it evaporate away.

THE INVITATION

Jesus is the Way and the Truth.
If you are not thinking that Jesus Christ is the Son of God,
you have rejected faith from the Holy Spirit.

That is the how it works.
Period.

Here it is:
If you do nothing, it is not a No-Op
That is not the *default parameter*.

This flowchart was created by the Creator of the Universe
You can't fill it in the way you want it to read
In science we are used to the autonomy of our minds
The reality is the autonomy of our minds
is our free will that we have.
The true default parameter is that we will not have a release
from all pain and suffering in eternal life in heaven.
The true default parameter is an eternity in hell made up of
unimaginable pain and suffering,
so infinitely beyond what we have ever endured on Earth
as to be unrecognizable in its horror.

Has God worked through this book
to enable your choice?
Try it!
Override the default parameter!

FINE PRINT?

So, I am putting aside my mind?
My reasoning?
My analytical skills?
For what?
A fairy tale, as Stephen Hawking calls it?
These are the excuses of an Age that is passing:
where a huge reason to not believe discussed earlier
in this small book is now lifted.
See clearly past it.
Your science mind and science skill
are why this book was written.
God is missing his children like you in heaven.
God is creating a miracle of modern times to reach to you.
He has already given you your talent, your Earth,
the air to breathe,
the Universe to explore.
So what's the catch?
What do you have to do to achieve an eternity in a paradise
beyond what can even be fathomed on Earth?

Eternal Life.
Not so bad considering its paradise.
What might you learn?
What might you discover?
When you have forever to do it?

THE INVITATION

Don't make the mistake.
You are smarter than simply rejecting God.
If you think this is the most important subject you will ever
address upon this Earth, and
that none of us knows when our time on Earth is at hand
then the Holy Spirit will have given you faith in Jesus.
At some point, this test cannot be made up.
The dissertation process ends in failure.
You do not acquire the PhD.
No second chances.

If these things ring true to you,
Choose now to not reject Him!
Lift the heaviest burdens of your life.
Change inside. Change your heart.
Continue your research.
Continue the enormous blessings you have known in
your life and continue to receive.
Your work is important to God.
Your work in this world has been acknowledged by God,
by His desire to reach you, his child.
We each have our purpose in His eyes,
if we will only seek His will.
Are science minds needed in the
New Earth/New Creation/New Age?
God loves all his children.
You are his beloved child.
You are not alone.
Your sadness and suffering will be gone one day in heaven.
How profound is this miracle?
What power do I have?

I have been empowered to offer you
a reason to not reject God,
You have permission to reconsider the choice to do so.

Right here. Right now.
Would you want The Holy Spirit to try to begin
to give you faith to know
that Jesus Christ is the Son of God?
Is there anything in this little book
that sounds like truthfulness that
may resonate within you?

Is the amazing love of God seeking and sacrificing His Son
stirring just a spark of gratitude?
Could you seek to know the Holy Spirit better?

What is the best thing to do?
It is good idea to attend church for renewal and insight
should you receive trust in the saving ways of Jesus.
Satan is really good at pulling you away.
Yes, you!

Christians do good works.
What does that mean?
It means they did not remain hostile to God,
and when called by the Holy Spirit,
they have allowed Jesus to live through them.

Would you want the light of Jesus to shine through you
so that others may glorify your Father?
What if you can't imagine doing those kind of good works
and thus find a reason to reject God?
This isn't rocket science, but the Lord wants his
rocket scientist children
to come home to Him in the next world.

THE INVITATION

Fall to your knees – no one can come to a true relationship
with God the Father
without trusting in the saving works of his Son,
Jesus.
Go ahead and
confess to Him that you have been resisting Him
to this point.
Go ahead and say,
"Please forgive me, Lord."

Because He already has forgiven you.

You will find what you are seeking!

You will find that
He knows you better than you know yourself!

The Messenger is a Scientist

Despite a rich and meaningful life you have been granted on this Earth, every box or gift you have ever received on Earth eventually is empty. Eventually, you realize that you are alone and your burdens are often overwhelming and the gifts of this world fade away because this is not your home.

God has tried countless times in your life to tell you where the gifts of life, food, shelter, the success you have achieved, and the very air you breathe come from. None of this is to the exclusion of the science you know that explains a great many things! You have applied your intellect to God's creations and God Himself created science. The overwhelming vastness and scientific mysteries of the Universe are a testament to Him and his love of us! Because there is one more gift God wishes us to have. And neither you nor I nor anyone you have met deserves or has earned this gift. It is given with grace, forgiveness of our sins, and we for incomprehensible reasons are asked to receive it. God gave his only Son to this world to make the Word human flesh, so that we could be redeemed and our sins completely forgiven. Who has ever done such a thing for you in this world? Many in this world have given their lives for others. Who alone has given their only son for you?

At a very early age, a scientist who is like you endured unimaginable pain and suffering. In many near death experiences, I realized I had visited God's perfect heaven and seen many miracles. In recent months of remembering, I learned that God had a purpose for me, as he does for all of us. I check my understanding with the Bible. It is very clear to me that those who have not seen and yet believe, are blessed by God, for their knowledge of Him is attained as it should be through the Bible. I am amazed that with the guidance of my Pastor, the more I study and understand our church's interpretation of Scripture, the more grateful I am at how much God revealed of Himself through the Bible. Perhaps I survived my early childhood to bring a message of God's concern about his scientist children back to you so that you could have a glimpse of a picture that is even vaster than the Universe we explore. I want to share a glimpse of the majesty and the reality I saw and see now in the Bible, that it might penetrate your scientific mind to look even closer into what you have not yet explored, the very heart of the Universe.

There are many others who join with me to reach a peace between Christianity and science, on Earth. I am married to one such person. He has discerned the removal of the long-standing reason to not believe that gives scientists permission to reconsider God's grace and not let their rejection of God stand.

Dave and I are Missouri Synod Lutherans. I strongly believe that we have received a proper Christian education from our Pastor in the setting of our church. I am grateful and blessed for this!

If you apply your scientific mind beyond its field of expertise to places where it is not best used; if you decide that people like me – a fellow scientist, are delusional or misguided despite the countless miracles I saw, then you are putting in place another diabolical reason to not believe where the most important one has already fallen away. As that previous reason has served to do, it will only cause what it always has, if you, in effect reinstate it, by constantly supplying reasons from your mind.

The messengers are scientists like you. What do I have to gain by confessing Christ in this way? Only to shine like the stars, by leading you to righteousness, through the Lord's mercy and grace.

Cultural Exchange

You are a scientist. Open to what the Universe supplies evidence for. You have even wondered if there is life on other worlds. You acknowledge the possibility of life on other worlds. You would even likely accept evidence of life on other worlds, more intelligent than we. They too would have a Creator. The Trinity of the Father, the Son, and the Holy Spirit, three in one God, the highest level of existence in the Universe, the One who created it.

Missouri Synod Lutherans do not mention in their doctrinal interpretation of the Bible the resolution of the reason to not believe of which this book has spoken. However, Dave and I as scientists are thriving in the environment of this church, with this barrier to our profession as scientists resolved. Dave spoke with a Pastor of this denomination, now retired, long after Dave had arrived at the resolution of this barrier years before. This Pastor explained independently to Dave how he himself had arrived at it.

We believe people can thrive in this church environment in which we have thrived. We as scientists also understand the social phenomenon of cultural exchange. Lives improve in the most important way imaginable with this transition to God loving and God fearing people of Earth, who are diversely from all standings in life, and who have all found the Way the Truth the Life the Alpha the Omega the Morning Star that is eternal in Jesus Christ. Their lives will not end in death.

If your knee-jerk response is *not* to reject God, then perhaps you will come to God as I did. If the Holy Spirit gives you faith, you will be able to see that God has extended His hand to you. Reach toward Him! Take His hand. Through this age and the next.

Follow Him.
Learn about Him.
Give your problems to Him.
Keep yourself open to the Lord!

THE INVITATION
Fellowship

Pose any questions to:

fewarechosen@lindamorabito.com

To find a Missouri Synod Lutheran church visit

www.lcms.org

To learn more about the work of
David Meyer and Linda Morabito Meyer visit

www.lindamorabito.com

May God bless you always!

~ *Notes* ~

Credits
Cover Hubble Ultra Deep Field 2014: NASA, ESA, H. Teplitz and M. Rafelski
(IPAC/Caltech), A. Koekemoer (STScl), R. Windhorst (Arizona State University),
and Z. Levay (STScl)
Title page X-ray Ring around Crab Pulsar: NASA/CXC/SAO
Back cover Crab Pulsar Optical Survey data and X-ray:
NASA, Chandra X-ray Observatory, SAO, DSS
Photograph ♥KS Meyer Personal Collection
Cover Design David and Linda Meyer

www.ingramcontent.com/pod-product-compliance
Lightning Source LLC
Chambersburg PA
CBHW060634030426
42337CB00018B/3364